CLOUD CLEARING: IN YOUR OWN WORDS

Math Journal for Graduate Students

By

Amy Moore

First Edition

ISBN: 978-1-7331904-8-0

First Printing, 2021

Edited by editor_flora@fiverr.com.

The cover designer was PAYPERWORK LLC, eogbonna@payperworkllc.com

Preface

As a college student, I wrote in a journal to help me understand difficult concepts from my calculus course. I would use stacks of scrap paper to practice a math concept. I used to keep a journal of math concepts to give me confidence about a course. As you use this journal, please remember to use your required reading materials as the foundation for learning. I recommend reading the textbook for your math course before making journal entries in this book. I also recommend that one journaling session is no longer than 60 minutes, as breaks tend to help one retain information better. People have found it helpful to go over concepts with faculty or tutors- the journal can be shared with others as you wish. Please feel free to reach out to the author via email at amoore20@moorestat.com . You can also visit the author's shop at www.moorestat.com. This journal should help you to organize your thoughts about a topic. Never stop learning!

Example Journal Entry

Topic: _*Average of three Numbers*_

What is the specific Concept that is Confusing?

I do not understand how to get the average for three numbers.

Any Equations:

Mean(three numbers) = $\frac{X1 + X2 + X3}{Number\ Values\ in\ the\ Sample}$

Worked Out Sample Problem

Give 3 numbers : (20,20,8)

16 = $\frac{20 + 20 + 8}{3}$

Answer: 16

Topic: ______________________________

What is the specific Concept that is Confusing?

Any Equations:

Worked Out Sample Problem

Topic: ___________________________________

What is the specific Concept that is Confusing?

Any Equations:

Worked Out Sample Problem

Topic: ______________________________

What is the specific Concept that is Confusing?

Any Equations:

Worked Out Sample Problem

Topic: ______________________________

What is the specific Concept that is Confusing?

Any Equations:

Worked Out Sample Problem

Topic: ______________________________

What is the specific Concept that is Confusing?

Any Equations:

Worked Out Sample Problem

Topic: ______________________________

What is the specific Concept that is Confusing?

Any Equations:

Worked Out Sample Problem

Topic: ______________________________

What is the specific Concept that is Confusing?

Any Equations:

Worked Out Sample Problem

Topic: ________________________________

What is the specific Concept that is Confusing?

Any Equations:

Worked Out Sample Problem

Topic: ______________________________

What is the specific Concept that is Confusing?

Any Equations:

Worked Out Sample Problem

Topic: ______________________________

What is the specific Concept that is Confusing?

Any Equations:

Worked Out Sample Problem

Topic: ______________________________

What is the specific Concept that is Confusing?

Any Equations:

Worked Out Sample Problem

Topic: ____

What is the specific Concept that is Confusing?

Any Equations:

Worked Out Sample Problem

Topic: ___________________________________

What is the specific Concept that is Confusing?

Any Equations:

Worked Out Sample Problem

Topic: ______________________________

What is the specific Concept that is Confusing?

Any Equations:

Worked Out Sample Problem

Topic: ___________________________________

What is the specific Concept that is Confusing?

Any Equations:

Worked Out Sample Problem

Topic: ___________________________________

What is the specific Concept that is Confusing?

Any Equations:

Worked Out Sample Problem

Topic: ______________________________

What is the specific Concept that is Confusing?

Any Equations:

Worked Out Sample Problem

Topic: ________________________________

What is the specific Concept that is Confusing?

Any Equations:

Worked Out Sample Problem

Topic: ______________________________

What is the specific Concept that is Confusing?

Any Equations:

Worked Out Sample Problem

Topic: ____________________________________

What is the specific Concept that is Confusing?

Any Equations:

Worked Out Sample Problem

Topic: ______________________________

What is the specific Concept that is Confusing?

Any Equations:

Worked Out Sample Problem

Topic: ______________________________

What is the specific Concept that is Confusing?

Any Equations:

Worked Out Sample Problem

Topic: ___________________________________

What is the specific Concept that is Confusing?

Any Equations:

Worked Out Sample Problem

Topic: ______________________________

What is the specific Concept that is Confusing?

Any Equations:

Worked Out Sample Problem

Topic: ___________________________________

What is the specific Concept that is Confusing?

Any Equations:

Worked Out Sample Problem

Topic: ___________________________________

What is the specific Concept that is Confusing?

Any Equations:

Worked Out Sample Problem

Topic: ___________________________________

What is the specific Concept that is Confusing?

Any Equations:

Worked Out Sample Problem

Topic: ______________________________

What is the specific Concept that is Confusing?

Any Equations:

Worked Out Sample Problem

Topic: ________________________________

What is the specific Concept that is Confusing?

Any Equations:

Worked Out Sample Problem

Topic: ______________________________

What is the specific Concept that is Confusing?

Any Equations:

Worked Out Sample Problem

Topic: ______________________________

What is the specific Concept that is Confusing?

Any Equations:

Worked Out Sample Problem

Topic: ___________________________________

What is the specific Concept that is Confusing?

Any Equations:

Worked Out Sample Problem

Topic: ______________________________

What is the specific Concept that is Confusing?

Any Equations:

Worked Out Sample Problem

Topic: ______________________________

What is the specific Concept that is Confusing?

Any Equations:

Worked Out Sample Problem

Topic: ______________________________________

What is the specific Concept that is Confusing?

Any Equations:

Worked Out Sample Problem

Topic: ______________________________

What is the specific Concept that is Confusing?

Any Equations:

Worked Out Sample Problem

Topic: ________________________________

What is the specific Concept that is Confusing?

Any Equations:

Worked Out Sample Problem

Topic: ______________________________

What is the specific Concept that is Confusing?

Any Equations:

Worked Out Sample Problem

Topic: ______________________________

What is the specific Concept that is Confusing?

Any Equations:

Worked Out Sample Problem

Topic: ____________________________________

What is the specific Concept that is Confusing?

Any Equations:

Worked Out Sample Problem

Topic: ________________________________

What is the specific Concept that is Confusing?

Any Equations:

Worked Out Sample Problem

Topic: ___________________________________

What is the specific Concept that is Confusing?

Any Equations:

Worked Out Sample Problem

Topic: ______________________________

What is the specific Concept that is Confusing?

Any Equations:

Worked Out Sample Problem

Topic: ___________________________________

What is the specific Concept that is Confusing?

Any Equations:

Worked Out Sample Problem

Topic: ______________________________

What is the specific Concept that is Confusing?

Any Equations:

Worked Out Sample Problem

Topic: ______________________________

What is the specific Concept that is Confusing?

Any Equations:

Worked Out Sample Problem

Topic: ______________________________

What is the specific Concept that is Confusing?

Any Equations:

Worked Out Sample Problem

Topic: ________________________________

What is the specific Concept that is Confusing?

Any Equations:

Worked Out Sample Problem

Topic: ________________________________

What is the specific Concept that is Confusing?

Any Equations:

Worked Out Sample Problem

Topic: ______________________________

What is the specific Concept that is Confusing?

Any Equations:

Worked Out Sample Problem

Topic: ___________________________________

What is the specific Concept that is Confusing?

Any Equations:

Worked Out Sample Problem

Topic: ______________________________

What is the specific Concept that is Confusing?

Any Equations:

Worked Out Sample Problem

Topic: ______________________________

What is the specific Concept that is Confusing?

Any Equations:

Worked Out Sample Problem

Topic: ________________________________

What is the specific Concept that is Confusing?

Any Equations:

Worked Out Sample Problem

Topic: ____________________________________

What is the specific Concept that is Confusing?

Any Equations:

Worked Out Sample Problem

Topic: ________________________________

What is the specific Concept that is Confusing?

Any Equations:

Worked Out Sample Problem

Topic: ________________________________

What is the specific Concept that is Confusing?

Any Equations:

Worked Out Sample Problem

Topic: ______________________________

What is the specific Concept that is Confusing?

Any Equations:

Worked Out Sample Problem

Topic: ______________________________

What is the specific Concept that is Confusing?

Any Equations:

Worked Out Sample Problem

Topic: ______________________________

What is the specific Concept that is Confusing?

Any Equations:

Worked Out Sample Problem

Topic: ___________________________________

What is the specific Concept that is Confusing?

Any Equations:

Worked Out Sample Problem

Topic: ______________________________

What is the specific Concept that is Confusing?

Any Equations:

Worked Out Sample Problem

Topic: ___________________________________

What is the specific Concept that is Confusing?

Any Equations:

Worked Out Sample Problem

Topic: ______________________________

What is the specific Concept that is Confusing?

Any Equations:

Worked Out Sample Problem

Topic: ______________________________

What is the specific Concept that is Confusing?

Any Equations:

Worked Out Sample Problem

Topic: ___________________________________

What is the specific Concept that is Confusing?

Any Equations:

Worked Out Sample Problem

Topic: ___________________________________

What is the specific Concept that is Confusing?

Any Equations:

Worked Out Sample Problem

Topic: ______________________________

What is the specific Concept that is Confusing?

Any Equations:

Worked Out Sample Problem

Topic: ___________________________________

What is the specific Concept that is Confusing?

Any Equations:

Worked Out Sample Problem

Topic: ______________________________

What is the specific Concept that is Confusing?

Any Equations:

Worked Out Sample Problem

Topic: ______________________________

What is the specific Concept that is Confusing?

Any Equations:

Worked Out Sample Problem

Topic: ______________________________

What is the specific Concept that is Confusing?

Any Equations:

Worked Out Sample Problem

Topic: ___________________________________

What is the specific Concept that is Confusing?

Any Equations:

Worked Out Sample Problem

Topic: ______________________________

What is the specific Concept that is Confusing?

Any Equations:

Worked Out Sample Problem

Topic: ______________________________

What is the specific Concept that is Confusing?

Any Equations:

Worked Out Sample Problem

Topic: ________________________________

What is the specific Concept that is Confusing?

Any Equations:

Worked Out Sample Problem

Topic: ________________________________

What is the specific Concept that is Confusing?

Any Equations:

Worked Out Sample Problem

Topic: ______________________________

What is the specific Concept that is Confusing?

Any Equations:

Worked Out Sample Problem

Topic: ______________________________

What is the specific Concept that is Confusing?

Any Equations:

Worked Out Sample Problem

Topic: ______________________________

What is the specific Concept that is Confusing?

Any Equations:

Worked Out Sample Problem

Topic: ______________________________

What is the specific Concept that is Confusing?

Any Equations:

Worked Out Sample Problem

Topic: ______________________________

What is the specific Concept that is Confusing?

Any Equations:

Worked Out Sample Problem

Topic: ______________________________

What is the specific Concept that is Confusing?

Any Equations:

Worked Out Sample Problem

Topic: ______________________________

What is the specific Concept that is Confusing?

Any Equations:

Worked Out Sample Problem

Topic: ________________________________

What is the specific Concept that is Confusing?

Any Equations:

Worked Out Sample Problem

Topic: ________________________________

What is the specific Concept that is Confusing?

Any Equations:

Worked Out Sample Problem

Topic: ______________________________

What is the specific Concept that is Confusing?

Any Equations:

Worked Out Sample Problem

Topic: ___________________________________

What is the specific Concept that is Confusing?

Any Equations:

Worked Out Sample Problem

Topic: ___________________________________

What is the specific Concept that is Confusing?

Any Equations:

Worked Out Sample Problem

Topic: ______________________________

What is the specific Concept that is Confusing?

Any Equations:

Worked Out Sample Problem

Topic: ___________________________________

What is the specific Concept that is Confusing?

Any Equations:

Worked Out Sample Problem

Topic: ______________________________

What is the specific Concept that is Confusing?

Any Equations:

Worked Out Sample Problem

Topic: ________________________________

What is the specific Concept that is Confusing?

Any Equations:

Worked Out Sample Problem

Topic: ____________________________________

What is the specific Concept that is Confusing?

Any Equations:

Worked Out Sample Problem

Topic: ______________________________

What is the specific Concept that is Confusing?

Any Equations:

Worked Out Sample Problem

Topic: ______________________________

What is the specific Concept that is Confusing?

Any Equations:

Worked Out Sample Problem

Topic: ___________________________________

What is the specific Concept that is Confusing?

Any Equations:

Worked Out Sample Problem

Topic: ______________________________

What is the specific Concept that is Confusing?

Any Equations:

Worked Out Sample Problem

Topic: ______________________________

What is the specific Concept that is Confusing?

Any Equations:

Worked Out Sample Problem

Topic: ________________________________

What is the specific Concept that is Confusing?

Any Equations:

Worked Out Sample Problem

Topic: ______________________________

What is the specific Concept that is Confusing?

Any Equations:

Worked Out Sample Problem

Topic: ______________________________

What is the specific Concept that is Confusing?

Any Equations:

Worked Out Sample Problem

Topic: ______________________________

What is the specific Concept that is Confusing?

Any Equations:

Worked Out Sample Problem

Topic: ______________________________

What is the specific Concept that is Confusing?

Any Equations:

Worked Out Sample Problem

Topic: ________________________________

What is the specific Concept that is Confusing?

Any Equations:

Worked Out Sample Problem

Topic: ______________________________

What is the specific Concept that is Confusing?

Any Equations:

Worked Out Sample Problem

Topic: ________________________________

What is the specific Concept that is Confusing?

Any Equations:

Worked Out Sample Problem

Topic: ____________________________

What is the specific Concept that is Confusing?

Any Equations:

Worked Out Sample Problem

Topic: ________________________________

What is the specific Concept that is Confusing?

Any Equations:

Worked Out Sample Problem

Topic: ________________________________

What is the specific Concept that is Confusing?

Any Equations:

Worked Out Sample Problem

Topic: ___________________________________

What is the specific Concept that is Confusing?

Any Equations:

Worked Out Sample Problem

Topic: ___________________________________

What is the specific Concept that is Confusing?

Any Equations:

Worked Out Sample Problem

Topic: ______________________________

What is the specific Concept that is Confusing?

Any Equations:

Worked Out Sample Problem

Topic: ______________________________

What is the specific Concept that is Confusing?

Any Equations:

Worked Out Sample Problem

Topic: ______________________________

What is the specific Concept that is Confusing?

Any Equations:

Worked Out Sample Problem

Topic: ______________________________

What is the specific Concept that is Confusing?

Any Equations:

Worked Out Sample Problem

Topic: ___________________________________

What is the specific Concept that is Confusing?

Any Equations:

Worked Out Sample Problem

Topic: ______________________________

What is the specific Concept that is Confusing?

Any Equations:

Worked Out Sample Problem

Topic: ___________________________________

What is the specific Concept that is Confusing?

Any Equations:

Worked Out Sample Problem

Topic: ______________________________

What is the specific Concept that is Confusing?

Any Equations:

Worked Out Sample Problem

Topic: ______________________________

What is the specific Concept that is Confusing?

Any Equations:

Worked Out Sample Problem

Topic: ________________________________

What is the specific Concept that is Confusing?

Any Equations:

Worked Out Sample Problem

Topic: ______________________________

What is the specific Concept that is Confusing?

Any Equations:

Worked Out Sample Problem

Topic: ________________________________

What is the specific Concept that is Confusing?

Any Equations:

Worked Out Sample Problem

Topic: ______________________________

What is the specific Concept that is Confusing?

Any Equations:

Worked Out Sample Problem

Topic: ______________________________

What is the specific Concept that is Confusing?

Any Equations:

Worked Out Sample Problem

Topic: ____________________________________

What is the specific Concept that is Confusing?

Any Equations:

Worked Out Sample Problem

Topic: ______________________________

What is the specific Concept that is Confusing?

Any Equations:

Worked Out Sample Problem

Topic: ________________________________

What is the specific Concept that is Confusing?

Any Equations:

Worked Out Sample Problem

Topic: ______________________________

What is the specific Concept that is Confusing?

Any Equations:

Worked Out Sample Problem

Topic: ______________________________

What is the specific Concept that is Confusing?

Any Equations:

Worked Out Sample Problem

Topic: ______________________________

What is the specific Concept that is Confusing?

Any Equations:

Worked Out Sample Problem

Topic: ______________________________

What is the specific Concept that is Confusing?

Any Equations:

Worked Out Sample Problem

Topic: ______________________________

What is the specific Concept that is Confusing?

Any Equations:

Worked Out Sample Problem

Topic: ______________________________

What is the specific Concept that is Confusing?

Any Equations:

Worked Out Sample Problem

Topic: ______________________________

What is the specific Concept that is Confusing?

Any Equations:

Worked Out Sample Problem

Topic: ______________________________

What is the specific Concept that is Confusing?

Any Equations:

Worked Out Sample Problem

Topic: ________________________________

What is the specific Concept that is Confusing?

Any Equations:

Worked Out Sample Problem

Topic: ______________________________

What is the specific Concept that is Confusing?

Any Equations:

Worked Out Sample Problem

Topic: ______________________________

What is the specific Concept that is Confusing?

Any Equations:

Worked Out Sample Problem

Topic: ______________________________

What is the specific Concept that is Confusing?

Any Equations:

Worked Out Sample Problem

Topic: ________________________________

What is the specific Concept that is Confusing?

Any Equations:

Worked Out Sample Problem

Topic: ___________________________________

What is the specific Concept that is Confusing?

Any Equations:

Worked Out Sample Problem

Topic: ______________________________

What is the specific Concept that is Confusing?

Any Equations:

Worked Out Sample Problem

Topic: ___________________________________

What is the specific Concept that is Confusing?

Any Equations:

Worked Out Sample Problem

Topic: ______________________________

What is the specific Concept that is Confusing?

Any Equations:

Worked Out Sample Problem

Topic: ______________________________

What is the specific Concept that is Confusing?

Any Equations:

Worked Out Sample Problem

Topic: ______________________________

What is the specific Concept that is Confusing?

Any Equations:

Worked Out Sample Problem

Topic: ________________________________

What is the specific Concept that is Confusing?

Any Equations:

Worked Out Sample Problem

Topic: ______________________________

What is the specific Concept that is Confusing?

Any Equations:

Worked Out Sample Problem

Topic: ________________________________

What is the specific Concept that is Confusing?

Any Equations:

Worked Out Sample Problem

Topic: ______________________________

What is the specific Concept that is Confusing?

Any Equations:

Worked Out Sample Problem

Topic: ________________________________

What is the specific Concept that is Confusing?

Any Equations:

Worked Out Sample Problem

Topic: ______________________________

What is the specific Concept that is Confusing?

Any Equations:

Worked Out Sample Problem

Topic: ______________________________________

What is the specific Concept that is Confusing?

Any Equations:

Worked Out Sample Problem

Topic: ______________________________

What is the specific Concept that is Confusing?

Any Equations:

Worked Out Sample Problem

Topic: ________________________________

What is the specific Concept that is Confusing?

Any Equations:

Worked Out Sample Problem

Topic: ______________________________

What is the specific Concept that is Confusing?

Any Equations:

Worked Out Sample Problem

Topic: ______________________________

What is the specific Concept that is Confusing?

Any Equations:

Worked Out Sample Problem

Topic: ______________________________

What is the specific Concept that is Confusing?

Any Equations:

Worked Out Sample Problem

Made in the USA
Monee, IL
08 November 2026

67944873R00092